MW00634872

THE PASSION OF
SS. PERPETUA &
FELICITY, MM.
TOGETHER WITH
THE SERMONS OF
S. AUGUSTINE ON
THESE SAINTS

THE PASSION OF

SS.

PERPETUA AND FELICITY

MM.

A NEW EDITION AND TRANSLATION
OF THE LATIN TEXT

TOGETHER WITH THE

SERMONS OF S. AUGUSTINE

UPON THESE SAINTS

NOW FIRST TRANSLATED INTO ENGLISH

BY

W. H. SHEWRING

LONDON
SHEED AND WARD
MCMXXXI

NIHIL OBSTAT: EDUARDUS J. MAHONEY, S.TH.D.
CENSOR DEPUTATUS
IMPRIMATUR: EDM. CAN. SURMONT
VIC. GEN.
WESTMONASTERII, DIE XIIIA JULII, MCMXXXI

FIRST PUBLISHED SEPTEMBER, 1931
BY SHEED & WARD
FROM 31 PATERNOSTER ROW
LONDON, E.C. 4

PRINTED IN GREAT BRITAIN BY THE
WHITEFRIARS PRESS, LTD., LONDON
AND TONBRIDGE

AD
INSTRUMENTUM
ECCLESIAE

PREFACE

SOME five years ago I was asked by Mr. Stanley Morison to make a translation of the *Passion of S. Perpetua*, to be inset in his journal *The Fleuron ;* the book was to be illustrated by Mr. Eric Gill and printed from types designed by the artist. I made the translation—from the Cambridge text of Dr. Armitage Robinson—and after much unavoidable delay it was published at the end of last year in the final number of *The Fleuron.*

In the interval my interest in the *Passion* had grown. The fact that all separate and annotated editions of it were out of print, the accident that after the publication of the Cambridge text a new MS. had been found and discoveries had been made at Carthage, the reflection that even the excellent text of Franchi was not beyond slight improvement—suggested the idea of a small book containing a revised Latin text, a revised edition of my translation and an introduction embodying the research of specialists.

Hence comes the present work, which addresses two classes of readers. Scholars may be glad to have a separate Latin text of the *Passion*, without full collations but with sufficient critical notes to allow them the use of their judgment on points of importance ; while the translation and introduction may enable more general readers to acquaint themselves with the authentic record of one of the earliest and most famous of Christian martyrdoms.

I record my debt : to the books of many writers —above all those of Allard, Leclercq and Robinson —on whose wide erudition my work so often depends ; to Lady Helen Asquith, who read through my book in proof ; and to Mr. Stanley Morison, who permits my translation to be reprinted and to whom I owe the first impulse to a study which has been of great interest to myself and may perhaps be of use to others.

W. H. S.

May, 1931.

CONTENTS

INTRODUCTION

INTRODUCTION

I.—HISTORY OF THE MARTYRS

THE martyrs Perpetua and Felicity with their companions suffered in the amphitheatre of Carthage on the seventh of March, A.D. 203. They were natives, most probably of Carthage itself, perhaps of the neighbouring town of Thuburbo; and were arrested under the edict of Septimius Severus, promulgated in 202. The third century was not an era of general persecution; for after the first fury of Nero's reign the emperors, though without abrogating his decrees, had often refrained from enforcing them. The arrest of Christians was frequent, but it was irregular; and as the number of the faithful increased the magistrates gave up the hope of exterminating the new religion, and endeavoured only to check its course by the sacrifice for example's sake of a few from among its multitudes. The edict of Severus was aimed, it seems, not at families already Christian, but against new converts from paganism, and even of these many must have been spared (the Christian mother and brothers of S. Perpetua were not arrested); but many also were denounced and condemned.

Among these the most famous names are those of S. Perpetua and her comrades. Herself a young matron of the illustrious family of the Vibii, she

xiii

was doubtless chosen on that account as a conspicuous and exalted victim. Perhaps of her household were the slaves Felicity and Revocatus; of Secundulus and Saturninus we know only that they were of the same group of catechumens; while Saturus had been the means of their conversion and of his own will joined the other prisoners after their arrest.

By a not uncommon custom the martyrs were first placed in *custodia libera*—that is, they were guarded by sureties in a private house. Thence they were taken up to the State prison (it was built above the town) where they suffered not only the common rigours of imprisonment, but also (on one day at least) confinement to the stocks. Here S. Perpetua was visited by her pagan father and her Christian relatives; her husband, of whom we hear nothing, may have been a timid Christian and in hiding; more probably he had died before.

The sufferings of the martyrs in prison and the visions that consoled them, their trial, condemnation and martyrdom are related in the *Passion* itself with a force and directness which would render vain any repetition of the story here.[1] On some passages of special interest or difficulty I have added notes at the foot of my translation; but there remain a few points on which I should like to comment.

[1] The English reader may find it interesting to compare with this the Passion of the Martyrs of Lyons (A.D. 177) in the abridged but very beautiful translation of Walter Pater (*Marius the Epicurean*, ch. 26).

INTRODUCTION

(1) S. Perpetua's vision of her brother in Purgatory and of his release through her intercession has particular importance as a very early testimony to the Church's doctrine on the subject. As S. Augustine explains (*De Anima*, I. 10), the boy was no doubt a Christian, but was old enough to have committed venial sin after his baptism.

(2) The mention of the *Porta Sanavivaria* or *Gate of Life* in Chapters X. and XX. seems to call for fuller explanation than could conveniently be given in a footnote. This gate, placed at one end of the arena, was so called because through it, in the common contests between gladiators, the victorious fighters returned. At the opposite end was the *Porta Libitinensis* or *Gate of Death*, through which were carried the bodies of the slain. Hence in her vision S. Perpetua returns to the Gate of Life after her spiritual victory over the Devil. But in the martyrdom itself she and S. Felicity are called back from the contest to this gate, not as conquerors (for there was no question of their lives being spared), but as having now suffered enough from the beasts to earn a reprieve before they were finally beheaded.

(3) Some readers of the *Passion* have allowed themselves to be easily scandalised by the attitude of the martyrs to some of their persecutors. In Chapter XVII. Saturus reproaches the curiosity of the people visiting the captives on the eve of the martyrdom and threatens them with the judgment to come. And at the beginning of the contest itself (Chapter XVIII.) the martyrs say as they pass before

the procurator Hilarian : *Thou judgest us, but God shall judge thee.* Is this, it has been asked, the language of a Christian before his persecutors ? I would answer that though such severity might seem at first sight less fitting than the untroubled meekness of other martyrs—of the English Martyrs, for instance, praying for " Elizabeth Queene "—we must consider the circumstances of the Carthaginians and the intention with which they spoke. A martyr dies as a witness to the true faith, and it may be his duty towards his enemies to instruct them by his words of the faith which he is to seal with his death. Martyrs of a later time, whose slayers had once professed the religion they persecuted, had no need to speak of forsaken truths which were known already. For such men their blood was a sufficient testimony. But for Christians in the pagan world, whose religion for most of those without was a detested secret, the time of their arrest and martyrdom might often be the only occasion, as it was the last, for the public utterance of Christian doctrine. Some of them, we are told,[1] spoke to the judges at their trial of the Redemption, of eternal life, of the Blessed Sacrament ; they were not understood. More penetrating to their adversaries was the terrible admonition of the judgment of God, less difficult of apprehension and scarcely to be forgotten. This

---

[1] SS. Firmus and Rusticus ; SS. Irenaeus and Mustiola ; S. Terentianus. But the *Acts* of these martyrs are of doubtful authenticity.

admonition our martyrs made, and their clearest justification is that after the words of Saturus "many believed."

But the martyrs of Carthage were not the first who thought rebuke and warning no sin against charity. S. Stephen, first of martyrs, whose dying prayer was made for his persecutors, had used at his trial more bitter words than Saturus. *Ye stiff-necked and uncircumcised of heart and ears, ye do always resist the Holy Ghost . . . ye have been the slayers and murderers of the Just One.* And there is a more solemn example still. Our Lord Himself, arraigned before the High Priest, had prophesied the last coming upon the clouds of heaven before He spoke the words, *Father, forgive them, for they know not what they do.*

The bodies of the saints—how soon after the martyrdom we do not know—were laid in the *confessio* of the *Basilica Maiorum* at Carthage. Later they were removed, apparently to escape sacrilege from pagans violating the tombs of Christian churches and cemeteries; and they were never afterwards recovered. But in 1907 Father Delattre in his excavation of the basilica came upon the memorial stone which had once covered the bodies of the saints. In characters of the fourth century, the broken inscription runs :–

✠ (HIC) SVNT MARTY(RES)
✠ SATVRVS SATVRN(INVS)
✠ REVOCATVS SECV(NDVLVS)
✠ FELICIT PER(PE)T PAS (NON MART)

(Here are the martyrs Saturus, Saturninus, Revocatus, Secundulus, Felicity and Perpetua, who suffered on the Nones of March.)

Though their shrine perished early, their memory remained illustrious. A few years after the martyrdom, Tertullian in his *De Anima* (c. 55) had spoken eloquently of " the most valiant martyr Perpetua " and later S. Augustine celebrated the saints in the anniversary sermons translated in the present work. From the fourth century onwards their names have stood in the Roman Martyrology, and the Church has paid them continuous honour, not only upon their yearly feast (originally March 7th, now March 6th), but daily in their commemoration during the Canon of the Mass.[1] " To us also, Thy sinful servants, who trust in the multitude of Thy mercies, vouchsafe to give some part and fellowship with Thy holy Apostles and Martyrs; with John, Stephen, Matthias . . . Felicity, Perpetua, Agatha . . . and with all Thy Saints."

II.—TEXT AND AUTHORSHIP OF THE " PASSION "

The *Passion* itself falls easily into four parts; a short preface by some person unknown proclaiming the recent operations of the Holy Spirit and followed by a list of the martyrs; then the narrative of S. Perpetua's imprisonment and her four visions,

[1] Felicity is there named before Perpetua as being the first to consummate her martyrdom. (But some scholars hold that the Felicity commemorated in the Mass is not the Carthaginian martyr but the Roman *Felicitas Septem Filiorum*. See Burkitt, *St. Felicity in the Roman Mass, J. T. S.*, April, 1931.)

written ostensibly by herself; then the single version of Saturus, also given as his own narrative; and finally an account of the martyrdom with a peroration in the same strain as the preface, and apparently from the same hand. There exist two versions of the whole work, one in Latin and one in Greek. Thus we are faced immediately by three questions. In what language was the *Passion* originally written? Are the narratives of the martyrs authentic? Who is the nameless editor or redactor of the whole?

The first two questions are best answered together. In the Latin text there are clear differences of style between the parts given to Perpetua, Saturus, and the redactor. In the Greek no such differences are obvious, and the style is uniform. The retention of these characteristic distinctions in the Latin suggests two things; that it is the original; and that the redactor has not re-handled the narratives of the saints. Further investigation confirms this view. Dr. Armitage Robinson has shown that the Greek version is in general the more diffuse of the two, that it expands some difficult phrases of the Latin and omits others, and that it fails to render the play of words which occurs occasionally in the Latin—all marks which betray the translator. To the general evidences of style I have been able to add a detailed rhythmical analysis of the prose of the texts—a method of proof too technical for discussion here, but set out at length in two articles to which I would refer any interested reader (*Prose Rhythm in the Passio S'. Perpetuae, J. T. S.*, October, 1928 ; and *En Marge de la Passion des SS. P. et F., Rev. Bén.,* January, 193). Briefly, I think one may say with some confidence that the Latin text is the original,[1] and that the narratives of Perpetua and Saturus are, as they claim to be; the work of the martyrs themselves.

The redactor of the *Passion* is unnamed in the MSS. Of his identity there is no absolute proof, but such evidence as we have is strongly in favour of Tertullian. General features of style, the use of particular words, some peculiarities in Scriptural quotation, have their closest parallel in his authentic works ; the visions of Perpetua were familiar to him (*De An.*, c. 55) ; and if we may suppose him to have composed his preface and epilogue immediately after the martyrdom, the Montanistic tone observable in them may well foreshadow his lapse into heresy some four years later. Here again it is to Dr. Armitage Robinson that we owe the clearest and most detailed statement of the case (*The Passion of S. Perpetua*, pp. 47–58) ; his cumulative argument is extremely forcible, and his conclusion is accepted by many scholars, including M. de Labriolle.[2]

---

[1] I would not suggest that the matter is beyond discussion ; for the opposite view has been forcibly defended by Salonius, and is supported by the great authority of Professor A. C. Clark. But Professor Rendel Harris, who at first claimed the Greek as the original, has since taken the opposite side ; and in favour of the Latin are Batiffol, Duchesne, Franchi, de Labriolle, Leclercq and Robinson.

[2] Who adds further parallels from Tertullian's writings (*Crise Montaniste*, pp. 345–351). On the other side, Professor A. C.

If the redactor of the *Passion* was indeed so famous a writer as Tertullian, why is it that the MSS. give no indication of authorship? As far as I know, no explanation has hitherto been offered, but I think a reasonable one is not difficult to find. Ancient books intended for general publication naturally bore the author's name upon them: but it often happened that a private copy was made for a particular patron, and in such cases a title might sometimes be thought unnecessary. MSS. descended from such private copies would also lack a title; and it is probably for this reason that some important classical works both in Greek and in Latin remain anonymous (F. W. Hall, *A Companion to Classical Texts*, p. 14). A like fate may have befallen the *Passion*. Intended not for private but for liturgical use—it would be read in church upon the feast of the Saints (S. Augustine, *Serm.* 280, 282)—it must have been thought to possess a devotional rather than a literary importance. Its interest centred not in the preface and epilogue, but in the story of the martyrs themselves. Hence the copies used by the churches may well have lacked the redactor's name, and though his identity may at first have been well enough known, such knowledge would fade; and in any case, when Tertullian fell into open heresy it was scarcely likely that his name should continue to be associated with that of the orthodox martyrs.

Clark has told me that he finds the examples collected by Robinson and de Labriolle insufficient for their purpose.

INTRODUCTION

For of the martyrs' orthodoxy there can be no question. Tertullian nowhere claims them as Montanists, though had they been so, he would have been only too eager to use their authority in support of the new doctrines. In the narratives and utterances of the saints there is nothing on which a charge of heresy could be grounded ; and the only external influences perceptible in their words or thought are those of the Bible, of the *Apocalypse of Peter* and of the *Shepherd* of Hermas, a work which enjoyed at that time an authority almost Scriptural, and to which, incidentally, Tertullian as a Montanist took violent exception (*De Pudicitia*, X., 11 ; XX., 2). But indeed the memory of the martyrs may be perfectly vindicated by a single argument ; that from the time of their martyrdom they have been venerated as saints by the Catholic Church, an honour accorded always only to those who have died in communion with her.

There is one passage in S. Perpetua's story which has been suspected of alteration in favour of Montanistic practice. Towards the end of Chapter IV. (the Eucharistic vision of the Good Shepherd) S. Perpetua says that she received from Our Lord's hands a morsel of the curd or cheese which He had from the milk—*de caseo quod mulgebat dedit mihi quasi buccellam*. It has been suggested by MM. Monceaux and Allard that the word *caseum*, curd or cheese, is not in accord with the verb *mulgere*, to milk ; that Perpetua perhaps wrote *lacte*, milk, and that this was altered to *caseo* by the redactor to suggest the rite of the Artotyrites, a small sect of possibly Montanistic origin who used cheese instead of wine in the celebration of their Eucharist. The hypothesis is unnecessary. Ancient authorities say little of the Artotyrites, and it is not certain that they were Montanists at all. Further, the textual explanation raises more difficulties than it solves. The expression " to milk curd " (in the sense of making curd from milk) may be unusual, but it is intelligible ; whereas if Perpetua had been speaking of milk she would scarcely have added, as she does, " I received it *with joined hands* and *ate* it (*manducavi*)." It is true (though Monceaux and Allard do not mention this) that in the fourth of the sermons on the martyrs attributed to S. Augustine there comes the phrase " Perpetua received new milk from the sweet Shepherd," but in this allegorical sermon the author departs more than once not only from Perpetua's probable intention but from actual details of the visions (see pp. 58, 59).[1] The true explanation of the passage is in fact the simplest. Since Our Lord is spoken of under the figure of the Good Shepherd, it is natural that the food He gives to the faithful should be spoken of, by a similar figure, as " a morsel of curd."

III.—MSS. AND EDITIONS

The MSS. of the *Passion* now extant are four.

(1) A = Codex Casinensis (Bibl. Cas., IV., 204). Saec. XI.

---

[1] So too the Pseudo-Augustine *De Tempore Barbarico*, c. v., has *buccellam lactis ;* but he also is no trustworthy authority.

(2) B = Codex Compendiensis (Bibl. Nationale, Lat. 17626). Saec. X.

(3) M = Codex Mediolanensis vel Ambrosianus (Amb. C. 210 inf.). Saec. XI. exeunte.

(4) g = Codex Graecus (S. Sepulchri 1). Saec. X.

To these must be added (5) C = Codex Salisburgensis vel Sarisburiensis. This MS. is now lost, but is cited in the editions of Thomas Spark (1680) and of Ruinart (1689).

During the Middle Ages all MSS. of the authentic *Passion* had disappeared, and their place had been taken by the abbreviated and corrupted Latin *Acts* of the Martyrs, to which I shall not again refer. [Text in Robinson, pp. 100–103 ; discussion, *ibid.*, pp. 15–22, and in Monceaux (*Hist. litt. de l'Afrique chrétienne*, I., 79) and Allard (*Histoire des Persécutions*, II., 103–4).] Of the five MSS. above named, A was first recovered by Holsten at Monte Cassino (where it still is) ; he prepared an edition, but died before its completion, and the text was first published by Poussin at Rome in 1663. In 1680 at Oxford, Thomas Spark, of Christ Church, published an edition of the *Passion* together with Lactantius *De Mortibus Persecutorum ;* in this he noted some variants from a Codex Sarisburiensis (C). Ruinart in his edition of 1689 used what was undoubtedly the same MS. (C), though he calls it not Sarisburiensis, but Salisburgensis, together with B (Compendiensis), which is now at Paris. C, whether found originally at Salisbury or at Salzburg, cannot now be traced, and is known to us only by the quotations of Spark and Ruinart. In 1889 the Greek version (g) was discovered by Professor Rendel Harris at the Convent of the Holy Sepulchre in Jerusalem, and by him and Professor Gifford was published at Cambridge in 1890. Finally, in 1892 a new MS. (M) was found in the Ambrosian Library, and a collation of it printed in the same year in *Analecta Bollandiana* (pp. 100–102 ; 369–373).

The relation of the MSS. may be summarised as follows. A is an independent authority, often giving a correct reading where BCMg together are wrong, and a false one where BCMg together are right. BCMg form a related group ; from their common ancestor, g probably descends directly and BCM through the intermediary of a MS. in which the Preface was lacking—either through the accidental loss of a page or perhaps through deliberate omission due to some suspicion of its theological implications. Of the Latin MSS. of this group it is M which approaches nearest to the text of g.

None of these authorities is without value for the reconstruction of the text. The most accurate version is that of A, while M abounds in all manner of illiteracies. Yet in Chapter XX., M alone of the Latin MSS., preserves the essential *acu* before *requisita*, though its spelling of the latter word as *recusuta* shows how transient at best its virtue is. Of most texts, probably, the best edition must be eclectic ; with the *Passio* the necessity is clear.

The printed editions of the text are :—

Poussin, Rome, 1663. (*Editio princeps,* containing notes by Poussin and Holsten.)

(Thomas Spark), Oxford, 1680. (With variants from C.)

Ruinart, Rome, 1689. (In *Acta Sincera Martyrum,* p. 85 ; uses B and C, but with faulty collations ; reprinted in Migne, P.L. III., pp. 13–58.)

Harris and Gifford, Cambridge, 1890. (*Editio princeps* of the Greek, with Ruinart's Latin text revised.)

Armitage Robinson, Cambridge, 1891. (Latin and Greek text, with valuable introduction and notes and full collation of ABCg ; unfortunately M was unknown to the editor.)

Franchi de' Cavalieri, Rome, 1896. (5th supplement of *Römische Quartalschrift.* Latin and Greek. Admirably introduced, and including readings of M.)

Gebhardt, Berlin, 1902. (*Ausgewählte Märtyrerakten,* p. 61. Latin and Greek. The preface (p. vii.) refers to some " hitherto unutilised MSS.," details of which are promised for later publication. The author's death in 1920 left the MSS. unidentified, but he seems to have drawn almost nothing from them beyond the addition of one word in Chapter IV. Their value therefore seems to have been slight, and the apparatus gives one no clue to Gebhardt's use of them. His method is doubtful, since in some twenty places he assigns to

Franchi's conjecture what are actually well attested MS. readings.)

Knopf, Tübingen. (*Ausgewählte Märtyrerakten*, p. 35. First published in 1901; reprinted in Krüger's revision in 1929. Latin only. Has defective critical apparatus.)

Of the many general studies of the *Passion* (with or without translation) the most important are those of Allard (*Histoire des Persécutions*, II., 102–136) and of Leclercq (*Les Martyrs*, I. ; introduction and translation, pp. 120–139 ; and an invaluable general preface on the history and procedure of early martyrdoms, pp. vii.–cxvi.). Leclercq gives also a detailed bibliography (*ibid.*, pp. 121–122) to which the following references may be added : P. de Labriolle, *Hist. de la Litt. Lat. Chrétienne*, pp. 141–144, *La Crise Montaniste*, pp. 338–353, *Sources de l'Hist. du Montanisme*, pp. 9–11 (Paris, 1913–1924) ; Milman, *Hist. of Christianity*, II., pp. 165–172 (London, 1867) ; Monceaux, *Hist. Litt. de l'Afrique Chrétienne*, I., pp. 70–96, and *La Vraie Légende Dorée* (Paris, 1901 and 1928) ; Morin, *S. Augustini Tractatus Inediti*, pp. xxvi.–xxvii., 196–200, 210–211 (Zürich, 1918) ; E. C. E. Owen, *Some Authentic Acts of the Early Martyrs* (Oxford, 1927) ; Salonius, *Passio S. Perpetuae* (Helsingfors, 1918) ; Wallis, *Ante-Nicene Library*, XIII. (S. Cyprian II., pp. 276–292, Edinburgh, 1869 ; the first complete English translation) ; and finally *Dict. d'Archéologie Chrétienne*, *s.v. Carthage*.

INTRODUCTION

The text of the *Passion* is short ; it has already been edited by capable scholars ; and a new edition will not be expected to introduce many novelties. Apart from a few original emendations, my text differs from those of Robinson and Franchi only in the greater consideration given to the readings of BCMg ; and whether in conjecture or in the choice between variants my intention has chiefly been to do justice to three elementary things : the ordinary claims of grammar, the style of the three writers, and the sense of each passage as determined by the context. My decisions on particular points have been defended and explained in the two articles to which I have already referred (p. xx.).

My critical apparatus is limited, as in a work of this kind it was bound to be ; but I hope it is nowhere misleading. I have, of course, noted all conjectures and assigned them to their earliest authors. (The conjecture *dulce* in Chapter IV. was once made by myself (R. B., January, 1931) ; but I found later that I had been anticipated by Gebhardt, and the emendation is here duly attributed to him.[1]) Of MS. variants I note only (1) those which seem in themselves probable or possible alternatives to the reading chosen, and (2) those which provide a palæological basis for emendation.

---

[1] It may perhaps be the reading of one of his unpublished MSS. ; so too with *cultu* in c. VII.

The apparatus itself has been made as brief as possible. Reference by numerals avoids the repetition in the note of the reading given in the text; and where g gives the obvious equivalent of a particular Latin reading, it is classed with the Latin MSS. without verbal quotation. Thus in Chapter III. there is a choice to be made between *profectus* and *profecto* (adv.); here g supports the former word by reading ἐξῆλθεν, and my note accordingly gives " [profectus] *BCMg;* profecto *A.*"

In the matter of spelling I have followed modern practice in giving unassimilated forms throughout, without of course any presumption that these forms were actually used by the writers.

V.—THE SERMONS OF S. AUGUSTINE

In the printed editions of S. Augustine are included four sermons in honour of SS. Perpetua and Felicity which were preached upon their feast. Three of these (*Serm*. 280, 281, 282, Migne *P.L.* xxxviii., 1280–1286) are certainly genuine. The fourth was discovered by Holsten in the Vatican Library, and is printed by Migne among " doubtful " sermons (*Serm*. 394, *P.L.* xxxix., 1715–1716). Whether it may in fact be genuine I lack the competence to say; but it is interesting in itself, and seemed well worth translation.

A fifth sermon on the same subject was recently found by Dom Morin, who in 1918 printed it with other unpublished sermons of S. Augustine and his

school (see p. xxvii.), ascribing it doubtfully to
Quodvultdeus. It is a poor piece of work, turgid
in style and with many infelicitous imitations of
S. Augustine ; and I have not translated it here.

The other sermons, to the best of my knowledge,
are translated now for the first time. To anyone
who has read the *Passion* they need no explanation.
They are characteristic of their author ; eloquent
and sometimes rhetorical (the repeated play on the
names of the Saints will not be to everyone's taste),
but written with a dignity well sustained and with
more than one passage of great theological depth.

The text from which I have translated this part
of my book is that of the Maurists (*S. Aug. Op.* v.,
pp. 1134-1138 and 1508, Paris, 1683) which Migne
reprints. No variants are given and the MSS. are
not clearly indicated ; but I can find no later or
better edition and have no access to MSS. or
collations of them. Hence, after making one
emendation, I have translated the text as it stands.
Throughout the book Biblical quotations are
italicised ; they are all easily recognisable, and I
have thought it better therefore not to burden the
page with notes of reference.

THE PASSION OF SS. PERPETUA & FELICITY MM.

PASSIO SS. PERPETUAE ET FELICITATIS

1. Si vetera fidei exempla, et Dei gratiam testificantia et ædificationem hominis operantia, propterea in litteris sunt digesta, ut lectione eorum quasi repræsentatione [1] rerum et Deus honoretur et homo confortetur, cur non et nova documenta æque utrique causæ convenientia et digerantur? Vel quia proinde et hæc vetera futura quandoque sunt et necessaria posteris, si in præsenti suo tempore minori deputantur auctoritati, propter præsumptam venerationem antiquitatis. Sed viderint qui unam virtutem Spiritus unius Sancti pro ætatibus iudicent temporum : cum maiora reputanda sunt novitiora quæque ut novissimiora, secundum exsuperationem gratiæ in ultima sæculi spatia decretam. *In novissimis* enim *diebus, dicit Dominus, effundam de Spiritu meo super omnem carnem, et prophetabunt filii filiæque eorum : et super servos et ancillas meas de meo Spiritu effundam : et iuvenes visiones videbunt, et senes somnia somniabunt.* Itaque et nos qui sicut prophetias ita et visiones novas pariter repromissas et agnoscimus et honoramus, ceterasque virtutes Spiritus Sancti ad instrumentum Ecclesiæ deputamus, cui et missus est idem omnia donativa administrans in omnibus

---

[1] *Harris ;* repensatione *A.*

prout [1] unicuique distribuit Dominus, necessario et digerimus et ad gloriam Dei lectione celebramus ; ut ne qua aut imbecillitas aut desperatio fidei apud veteres tantum æstimet gratiam divinitatis [2] conversatam, sive in [3] martyrum sive in revelationum dignatione : cum semper Deus operetur quæ repromisit, non credentibus in testimonium, credentibus in beneficium. Et nos itaque *quod audivimus et contrectavimus adnuntiamus* et *vobis*, fratres et filioli ; uti [4] et vos qui interfuistis rememoremini gloriæ Domini, et qui nunc cognoscitis per auditum communionem habeatis cum sanctis martyribus, et per illos cum Domino Iesu Christo, cui est claritas et honor in sæcula sæculorum. Amen.

2. Adprehensi sunt adulescentes catechumini, Revocatus et Felicitas conserva eius, Saturninus et Secundulus. Inter hos et Vibia Perpetua, honeste nata, liberaliter instituta, matronaliter nupta, habens patrem et matrem et fratres duos, alterum æque catechuminum, et filium infantem ad ubera ; erat autem ipsa annorum circiter viginti et duorum. Hæc ordinem totum martyrii sui iam hinc ipsa narrabit, [5] sicut conscriptum manu sua et suo sensu reliquit.

3. Cum adhuc, inquit, cum prosecutoribus essemus, [6] et me pater verbis evertere cupiret et

---

[1] *edd. ;* pro *A.*
[2] *edd. ;* divinatis *A ;* divinitus *Hort.*
[3] *edd. ;* om. *A.*
[4] *nos ;* ut hii *A ;* ut *BM.*
[5] *Spark ;* narravit *codd.*
[6] *BMg ;* essem *A.*

4

deicere pro sua adfectione perseveraret: Pater, inquam, vides, verbi gratia, vas hoc iacens, urceolum sive aliud? Et dixit: Video. Et ego dixi ei: Numquid alio nomine vocari potest quam quod est? Et ait: Non. Sic et ego aliud me dicere non possum nisi quod sum, Christiana. Tunc pater motus hoc verbo mittit se in me, ut oculos mihi erueret; sed vexavit tantum, et profectus[1] est victus cum argumentis diaboli. Tunc paucis diebus quod caruissem patrem, Domino gratias egi, et refrigeravi absentia illius. In ipso spatio paucorum dierum baptizati sumus; mihi autem Spiritus dictavit non aliud petendum ab aqua, nisi sufferentiam carnis. Post paucos dies recipimur in carcerem; et expavi, quia numquam experta eram tales tenebras. O diem asperum! Æstus validus turbarum beneficio, concussuræ militum. Novissime macerabar sollicitudine infantis ibi. Tunc Tertius et Pomponius, benedicti diaconi qui nobis ministrabant, constituerunt præmio uti[2] paucis horis emissi in meliorem locum carceris refrigeraremus. Tunc exeuntes de carcere universi sibi vacabant; ego infantem lactabam iam inedia defectum. Sollicita pro eo adloquebar matrem et confortabam fratrem, commendabam filium. Tabescebam ideo quod illos tabescere videram mei beneficio. Tales sollicitudines multis diebus passa sum: et usurpavi ut mecum infans in carcere maneret; et statim convalui et relevata sum a

---

[1] *BCMg;* profecto *A.*
[2] *nos;* ut hii *A;* ut *BM.*

labore et sollicitudine infantis, et factus est mihi carcer subito prætorium, ut ibi mallem essem quam alicubi.

4. Tunc dixit mihi frater meus : Domina soror, iam in magna dignatione es, tanta ut postules visionem et ostendatur tibi an passio sit an commeatus. Et ego quæ me sciebam fabulari cum Domino, cuius beneficio tanta experta eram, fidenter repromisi ei, dicens : Crastina die tibi renuntiabo. Et postulavi, et ostensum est mihi hoc. Video scalam æream [1] miræ magnitudinis, pertingentem usque ad cælum, et angustam, per quam non nisi singuli adscendere possent, et in lateribus scalæ omne genus ferramentorum infixum. Erant ibi gladii, lanceæ, hami, macheræ [2]; ut si quis neglegenter aut non sursum adtendens adscenderet, laniaretur et carnes eius inhærerent ferramentis. Et erat sub ipsa scala draco cubans miræ magnitudinis, qui adscendentibus insidias parabat,[3] et exterrebat ne adscenderent. Adscendit autem Saturus prior, qui postea se propter nos ultro tradiderat, quia ipse nos ædificaverat, et tunc cum adducti sumus præsens non fuerat. Et pervenit in caput scalæ, et convertit se et dixit : Perpetua, sustineo te : sed vide ne te mordeat draco ille. Et dixi ego : Non me nocebit in nomine Iesu Christi. Et desub ipsa scala, quasi timens me, lente eiecit caput ; et quasi primum gradum calcarem, calcavi illi caput. Et

---

[1] *CMg;* aureaṁ *B;* *om. A.*

[2] *add.* verruti (*pro* veruta) *codd. nescio qui quos laudat Gebhardt;* *add.* ὀβελίσκων *g.*

[3] *BCM;* praestabat *A.*

adscendi, et vidi spatium inmensum horti, et in medio sedentem hominem canum, in habitu pastoris, grandem, oves mulgentem : et circum-stantes candidati,[1] milia multa. Et levavit caput et adspexit me, et dixit mihi : Bene venisti, tegnon. Et clamavit me, et de caseo quod mulgebat dedit mihi quasi buccellam ; et ego accepi iunctis mani-bus, et manducavi : et universi circumstantes dixerunt Amen. Et ad sonum vocis experrecta sum, commanducans adhuc dulce [2] nescio quid. Et retuli statim fratri meo, et intelleximus passionem esse futuram ; et cœpimus nullam iam spem in sæculo habere.

5. Post paucos dies rumor cucurrit quod audi-remur. Supervenit autem et de civitate pater meus, consumptus tædio ; et adscendit ad me, ut me deiceret, dicens : Miserere, filia, canis meis ; miserere patri, si dignus sum a te pater vocari ; si his te manibus ad hunc florem ætatis provexi ; et [3] te proposui omnibus fratribus tuis : ne me dederis in dedecus hominum. Adspice fratres tuos ; adspice matrem tuam et materteram ; adspice filium tuum, qui post te vivere non poterit. Depone animos ; ne universos nos extermines : nemo enim nostrum libere loquetur, si tu aliquid fueris passa. Hæc dicebat quasi [4] pater pro sua pietate, basians mihi manus, et se ad meos pedes iactans ; et

---

[1] *A ;* circumstantes candidatos *BC ;* circumstantium candi-datorum *Mg.*
[2] *Gebhardt ;* dulci *B ;* dulcedine *M ;* dulcis *Ruinart ; om. A.*
[3] *Mg ;* si *AB.*
[4] *BMg ; om. A.*

lacrimis non filiam sed dominam me vocabat.[1] Et ego dolebam causam patris mei, quod solus de passione mea gavisurus non esset de toto genere meo ; et confortavi eum, dicens : Hoc fiet in illa catasta quod Deus voluerit : scito enim nos non in nostra esse potestate constitutos,[2] sed in Dei. Et recessit a me contristatus.

6. Alio die cum pranderemus, subito rapti sumus ut audiremur, et pervenimus ad forum. Rumor statim per vicinas fori partes cucurrit, et factus est populus inmensus. Adscendimus in catastam. Interrogati ceteri confessi sunt. Ventum est et ad me. Et adparuit pater ilico [3] cum filio meo ; et extraxit me de gradu, dicens : Supplica, miserere infanti. Et Hilarianus procurator, qui tunc loco proconsulis Minuci Timiniani defuncti ius gladii acceperat : Parce, inquit, canis patris tui ; parce infantiæ pueri ; fac sacrum pro salute Imperatorum. Et ego respondi : Non facio. Et Hilarianus dixit : Christiana es ? Et ego respondi : Christiana sum. Et cum staret pater ad me deiciendam, iussus est ab Hilariano deici, et virga percussus est.[4] Et doluit mihi casus patris mei, quasi ego fuissem percussa ; sic dolui pro senecta eius misera. Tunc nos universos pronuntiat, et damnat ad bestias ; et hilares descendimus ad carcerem. Tunc quia consueverat a me infans

---

[1] *BM ;* me n. f. nominabat s. d. *A.*

[2] *A ;* futuros *BMg.*

[3] ilico (*sc.* in eo ipso loco)*A ;* illic *M ;* ibi *B ;* ἐκεῖ *g.*

[4] *BCM ;* percussit *A ;* τῶν δορυφόρων τις ἐτύπτησεν αὐτόν *g ; forsan* percussit eum quidam ex lictoribus.

mammas accipere, et mecum in carcere manere, statim mitto ad patrem Pomponium diaconum, postulans infantem. Sed pater dare noluit. Et quomodo Deus voluit, neque ille amplius mammas desiderat, neque mihi fervorem fecerunt; ne sollicitudine infantis et dolore mammarum macerarer.

7. Post dies paucos, dum universi oramus, subito media oratione profecta est mihi vox, et nominavi Dinocraten; et obstipui quod numquam mihi in mentem venisset nisi tunc; et dolui commemorata casus eius. Et cognovi me statim dignam esse, et pro eo petere debere. Et cœpi de ipso orationem facere multum, et ingemiscere ad Dominum. Continuo ipsa nocte ostensum est mihi hoc. Video Dinocraten exeuntem de loco tenebroso, ubi et complures erant, æstuantem valde et sitientem, sordido cultu [1] et colore pallido; et vulnus in facie eius, quod cum moreretur habuit. Hic Dinocrates fuerat frater meus carnalis, annorum septem, qui per infirmitatem facie cancerata male obiit, ita ut mors eius odio fuerit omnibus hominibus. Pro hoc ergo orationem feceram; et inter me et illum grande erat diastema,[2] ita ut uterque ad invicem accedere non possemus. Erat deinde in ipso loco ubi Dinocrates erat piscina plena aqua, altiorem marginem habens quam erat statura pueri; et extendebat se Dinocrates quasi bibiturus. Ego

---

[1] *Gebhardt;* ἐσθῆτα ἔχοντα *g;* vultu *ABCM.*
[2] *Holsten, et hoc ipsum vocabulum habet g;* diadema *A;* idiantem *B;* dianten *C;* spatium *M.*

9

dolebam quod et piscina illa aquam haberet, et tamen propter altitudinem marginis bibiturus non esset. Et experrecta sum, et cognovi fratrem meum laborare. Sed fidebam me profuturam labori eius : et orabam pro eo omnibus diebus quousque transivimus in carcerem castrensem ; munere enim castrensi eramus pugnaturi ; natale tunc Getæ Cæsaris. Et feci pro illo orationem die et nocte gemens et lacrimans ut mihi donaretur.

8. Die quo in nervo mansimus, ostensum est mihi hoc. Video locum illum quem retro videram,[1] et Dinocraten mundo corpore, bene vestitum, refrigerantem ; et ubi erat vulnus, video cicatricem : et piscinam illam quam retro videram, submisso margine usque ad umbilicum pueri ; et aquam de ea trahebat decurrentem [2] sine cessatione : et super marginem fiala erat aurea, plena aqua ; et adcessit Dinocrates, et de ea bibere cœpit : quæ fiala non deficiebat. Et satiatus abscessit [3] de aqua ludere more infantium gaudens. Et experrecta sum. Tunc intellexi translatum eum esse de pœna.

9. Deinde post paucos dies Pudens, miles optio præpositus carceris, qui nos magnificare cœpit intellegens [4] magnam virtutem [5] esse in nobis, multos ad nos admittebat, ut et nos et illi invicem

---

[1] *add.* tenebrosum esse lucidum *B.*

[2] *Franchi ; om. ABM ;* ἔρρεεν δὲ ἐξ αὐτῆς ἀδιαλείπτως ὕδωρ *g ;* et aqua de ea cadebat *Gebhardt ;* aqua inde extrabat *Radermacher.*

[3] *Spark ;* accessit *ABCM ;* (ἤρξατο *g*).

[4] *BC ;* magnifice coepit intelligere, *postea* qui multos *A ;* magnifice contempserat, c. intelligere *Robinson in commentario, nescio an recte.*

[5] *Ag ; add.* Dei *BM.*

refrigeraremus. Ut autem proximavit dies muneris, intrat ad me pater meus consumptus tædio, et cœpit barbam suam evellere et in terram mittere, et prosternere se in faciem, et improperare annis suis, et dicere tanta verba quæ moverent universam creaturam. Ego dolebam pro infelici senecta eius.

10. Pridie quam pugnaremus, video in horomate hoc venisse Pomponium diaconum ad ostium carceris, et pulsare vehementer. Et exivi ad eum, et aperui ei : qui erat vestitus discinctam candidam, habens multiplices galliculas. Et dixit mihi : Perpetua, te exspectamus ; veni. Et tenuit mihi manum, et cœpimus ire per aspera loca et flexuosa. Vix tandem pervenimus anhelantes ad amphitheatrum, et induxit me in media arena. Et dixit mihi : Noli pavere [1] ; hic sum tecum, et conlaboro tecum. Et abiit. Et adspicio populum ingentem adtonitum. Et quia sciebam me ad bestias damnatam esse, mirabar quod non mitterentur mihi bestiæ. Et exivit quidam contra me Ægyptius fœdus specie cum adiutoribus suis pugnaturus mecum. Veniunt et ad me adulescentes decori, adiutores et favisores [2] mei. Et exspoliata sum, et facta sum masculus. Et cœperunt me favisores mei oleo defrigere, quomodo solent in agonem : et illum contra Ægyptium video in afa volutantem. Et exivit vir quidam miræ magnitudinis, ut etiam excederet fastigium amphitheatri, discinctus tuni-

---

[1] *A ;* expavescere *BM.*

[2] *Franchi ; quod quidem postea loci, ubi hoc iteratur vocabulum, bis habet A ; hic* fautores *ACM ;* factores *B.*

cam, habens et purpuram [1] inter duos clavos per medium pectus, habens et galliculas multiformes ex auro et argento factas : efferens virgam quasi lanista, et ramum viridem in quo erant mala aurea. Et petiit silentium, et dixit : Hic Ægyptius, si hanc vicerit, occidet illam gladio ; hæc autem, si hunc vicerit, accipiet ramum istum. Et recessit. Et adcessimus ad invicem, et cœpimus mittere pugnos. Ille mihi pedes adprehendere volebat ; ego autem illi calcibus faciem cædebam. Et sublata sum in ære, et cœpi eum sic cædere quasi terram non calcans. At ubi vidi moram fieri, iunxi manus, ut digitos in digitos mitterem. Et adprehendi illi caput, et cecidit in faciem ; et calcavi illi caput. Et cœpit populus clamare, et favisores mei psallere. Et adcessi ad lanistam, et accepi ramum. Et osculatus est me, et dixit mihi : Filia, pax tecum. Et cœpi ire cum gloria ad portam Sanavivariam. Et experrecta sum ; et intellexi me non ad bestias, sed contra diabolum esse pugnaturam : sed sciebam mihi esse victoriam. Hoc usque in pridie muneris egi : ipsius autem muneris actum, si quis voluerit, scribat.

11. Sed et Saturus benedictus hanc visionem suam edidit, quam ipse conscripsit. Passi, inquit, eramus, et eximus de carne, et cœpimus ferri a quattuor angelis in orientem, quorum manus nos non tangebat. Ibamus autem non supini sursum

---

[1] *nos ;* discinctam habens t. et p. *BC ;* discinctatus purpuram *A ;* discinctus purpura *M ;* διεζωσμένος ἐσθῆτα ἥτις εἶχεν οὐ μόνον ἐκ τῶν δύο ὤμων τὴν πορφύραν (ἀλλὰ καὶ ἀνὰ μέσον ἐπὶ τοῦ στηθοῦς) g.

versi, sed quasi mollem clivum adscendentes. Et
liberato primo mundo vidimus lucem inmensam;
et dixi Perpetuæ (erat enim hæc a latere meo) : Hoc
est quod nobis Dominus promittebat; percepimus
promissionem. Et dum gestamur ab ipsis quattuor
angelis, factum est nobis spatium grande, quod tale
fuit quasi viridarium, arbores habens rosæ et omne
genus flores. Altitudo arborum erat in modum
cupressi, quarum folia canebant [1] sine cessatione.
Ibi autem in viridario alii quattuor angeli fuerunt,
clariores ceteris : qui ubi viderunt nos honorem
nobis dederunt, et dixerunt ceteris angelis : Ecce
sunt, ecce sunt : cum admiratione. Et expave-
scentes quattuor illi angeli, qui gestabant nos,
deposuerunt nos : et pedibus nostris transivimus
stadium via lata. Ibi invenimus Iocundum et
Saturninum et Artaxium, qui eadem persecutione
passi [2] vivi arserunt; et Quintum, qui et ipse
martyr in carcere exierat; et quærebamus de illis
ubi essent ceteri. Ceteri angeli dixerunt nobis [3] :
Venite prius, introite, et salutate Dominum.

12. Et venimus prope locum, cuius loci parietes
tales erant quasi de luce ædificati; et ante ostium
loci illius angeli quattuor stabant, qui introeuntes
vestierunt stolas candidas. Et introivimus, et
audivimus vocem unitam dicentem : Agios, agios,
agios : sine cessatione. Et vidimus in eodem loco [4]

---

[1] *Robinson ;* cadebant *ABMg ;* ardebant *C.*
[2] *BM ; om. A.*
[3] *nos ;* ubi e. c. Et d. n. ceteri angeli *M ;* ubi essent. Cet. a. d.
n. *A ;* ubi e. ceteri. D. autem n. a. *B.*
[4] *AM ;* medio loci illius *BCg.*

sedentem quasi hominem canum, niveos habentem capillos, et vultu iuvenili ; cuius pedes non vidimus. Et in dextra et in sinistra eius seniores quattuor ; et post illos [1] seniores complures stabant. Et introeuntes cum admiratione stetimus ante thronum : et quattuor angeli sublevaverunt nos, et osculati sumus illum, et de manu sua traiecit nobis in faciem. Et ceteri seniores dixerunt nobis : Stemus. Et stetimus, et pacem fecimus. Et dixerunt nobis seniores : Ite et ludite. Et dixi Perpetuæ : Habes quod vis. Et dixit mihi : Deo gratias, ut quomodo in carne hilaris fui, hilarior sim et iam modo.[2]

13. Et exivimus, et vidimus ante fores Optatum episcopum ad dexteram, et Aspasium presbyterum doctorem ad sinistram, separatos et tristes. Et miserunt se ad pedes nobis, et dixerunt : Componite inter nos, quia existis, et sic nos reliquistis. Et diximus illis : Non tu es papa noster, et tu presbyter, ut vos ad pedes nobis mittatis ? Et moti sumus et complexi illos sumus. Et cœpit Perpetua græce cum illis loqui ; et segregavimus eos in viridarium sub arbore rosæ. Et dum loquimur cum eis, dixerunt illis angeli : Sinite illos refrigerent ; et si quas habetis inter vos dissensiones, dimittite vobis invicem. Et conturbaverunt eos. Et dixerunt Optato : Conrige plebem tuam ; quia sic ad te conveniunt quasi de circo redeuntes, et de factionibus certantes. Et sic nobis visum est quasi

---

[1] *add.* ceteri *AB ; om. Mg.*
[2] *nos ;* sim etiam m. *M ;* sum hic m. *A ;* (hilariorem) et hic *B ;* χαρῶ νῦν *g.*

vellent claudere portas. Et cœpimus illic multos fratres cognoscere, sed et martyras ; ubi universi [1] odore inenarrabili alebamur, qui nos satiabat. Tunc gaudens experrectus sum.

14. Hæ visiones insigniores ipsorum martyrum beatissimorum Saturi et Perpetuæ, quas ipsi conscripserunt. Secundulum vero Deus maturiore exitu de sæculo adhuc in carcere evocavit, non sine gratia, ut bestias lucraretur. Gladium tamen etsi non anima certe caro eius agnovit.

15. Circa Felicitatem vero, et illi gratia Domini eiusmodi contigit. Cum octo iam mensium ventrem haberet (nam prægnans fuerat adprehensa), instante spectaculi die, in magno erat luctu, ne propter hoc [2] differretur ; quia non licet prægnantes pœnæ repræsentari ; et ne inter alios postea sceleratos sanctum et innocentem sanguinem funderet. Sed et commartyres eius graviter contristabantur, ne tam bonam sociam quasi comitem [3] in via eiusdem spei derelinquerent. [4] Coniuncto itaque unito gemitu ad Dominum orationem fuderunt ante tertium diem muneris. Statim post orationem dolores invaserunt. Et cum pro naturali difficultate octavi mensis in partu laborans doleret, ait illi quidam ex ministris cataractariorum : Quæ sic modo doles, quid facies obiecta bestiis, quas con-

---

[1] *M ; om.* ubi *A ; om.* univ. *B.*

[2] *M ;* ventrem *AB* (-e *A*) *; om. g.*

[3] *add.* solam *ABM ; om. g ; vocabulum in margine additum (ad sensum sequentis verbi* [de]relinquerent *explendum) mox huc inrepsisse suspicor ; saltem post* comitem *positum neque omnino cum verbo illo nec nisi inepte cum nomine* comitem *coniungi potest.*

[4] *Bg ;* relinquerent *AM.*

tempsisti cum sacrificare noluisti? Et illa respondit: Modo ego patior quod patior; illic autem alius erit in me qui patietur pro me, quia et ego pro illo passura sum. Ita enixa est puellam, quam sibi quædam soror in filiam educavit.

16. Quoniam ergo permisit et permittendo voluit Spiritus Sanctus ordinem ipsius muneris conscribi, etsi indigni ad supplementum tantæ gloriæ descri- bendæ, tamen quasi mandatum sanctissimæ Per- petuæ, immo fideicommissum eius exsequimur, unum adicientes documentum de ipsius constantia et animi sublimitate; quæ tribuno castigatius eos tractante, quia[1] ex admonitionibus hominum vanissimorum verebatur ne subtraherentur de carcere incantationibus aliquibus magicis, in faciem ei respondit: Quid utique non permittis nobis refrigerare noxiis nobilissimis, Cæsaris scilicet, et natali eiusdem pugnaturis? Aut non tua gloria est, si pinguiores illo producamur? Horruit et erubuit tribunus; et ita iussit eos humanius haberi, ut fratribus eius et ceteris facultas fieret introeundi, et refrigerandi cum eis; iam et ipso optione carceris credente.

17. Pridie quoque cum illam cenam ultimam, quam liberam vocant, quantum in ipsis erat non cenam liberam sed agapen cenarent, eadem con- stantia ad populum verba ista iactabant, con- minantes iudicium Dei, contestantes passionis suæ

---

[1] *nos;* cum a trib. castigatius eo tractanti quia *A;* quia trib. castiganti (–e *C* apud Ruinart) eos et male tractante quoniam *BC;* qua trib. castiganti eos et male tractanti qui *M;* τοῦ χιλιάρχου ἀπηνέστερον αὐτοῖς προσφερομένου *g.*

felicitatem, inritantes [1] concurrentium curiositatem, dicente Saturo : Crastinus dies vobis satis non est ? Quid [2] libenter videtis quod odistis ? Hodie amici, cras inimici. Notate tamen vobis [3] facies nostras diligenter, ut recognoscatis nos in die illo. Ita omnes inde adtoniti discedebant ; ex quibus multi crediderunt.

18. Inluxit dies victoriæ illorum, et processerunt de carcere in amphitheatrum quasi in cælum ituri, hilares, vultu decori ; si forte, gaudio paventes, non timore. Sequebatur Perpetua lucido incessu, ut matrona Christi, ut Dei delicata, vigore oculorum deiciens omnium conspectum. Item Felicitas, salvam se peperisse gaudens ut ad bestias pugnaret, a sanguine ad sanguinem, ab obstetrice ad retiarium, lotura post partum baptismo secundo. Et cum deducti essent in portam, et cogerentur habitum induere, viri quidem sacerdotum Saturni, feminæ vero sacratarum Cereri, generosa illa in finem usque constantia repugnavit. Dicebat enim : Ideo ad hoc sponte pervenimus, ne libertas nostra obduceretur ; ideo animam nostram addiximus, ne tale aliquid faceremus : hoc vobiscum pacti sumus. Adgnovit iniustitia iustitiam : concessit tribunus, quomodo erant, simpliciter inducerentur. Perpetua psallebat, caput iam Ægyptii calcans. Revocatus et Saturninus et Saturus populo spectanti conminabantur. Dehinc ut sub conspectu Hilariani

[1] inritantes (*i.e.* pro nihilo ducentes) *A ;* inredeantes *B ;* irridentes *CM.*

[2] *CMg ;* qui *B ;* quod *A.*

[3] *BCM ;* nobis *A.*

17

pervenerunt, gestu et nutu cœperunt Hilariano dicere : Tu nos, inquiunt, te autem Deus. Ad hoc populus exasperatus flagellis vexari eos pro ordine venatorum postulavit. Et utique gratulati sunt, quod aliquid et de dominicis passionibus essent consecuti.

19. Sed qui dixerat *Petite et accipietis* petentibus dederat eum exitum quem quisque desideraverat. Nam si quando inter se de martyrii sui voto sermocinabantur, Saturninus quidem omnibus bestiis velle se obici profitebatur, ut scilicet gloriosiorem gestaret coronam. Itaque in conmissione spectaculi ipse cum Revocato leopardo expertus etiam super pulpitum ab urso erat vexatus.[1] Saturus autem nihil magis quam ursum abominabatur ; sed uno morsu leopardi confici se iam præsumebat. Itaque cum apro subministraretur, venator potius qui illum apro subligaverat, sub-fossus ab eadem bestia, post dies muneris obiit : Saturus solum modo tractus est. Et cum ad ursum substrictus esset in ponte, ursus de cavea prodire noluit. Itaque secundo Saturus inlæsus revocatur.

20. Puellis autem ferocissimam vaccam, ideoque præter consuetudinem comparatam, diabolus præparavit, sexui earum etiam de bestia æmulatus. Itaque dispoliatæ et reticulis indutæ producebantur. Horruit populus, alteram respiciens puellam delicatam, alteram a partu recentem stillantibus mam-

---

[1] *Franchi, et sic fere g ;* i. c. s. ipse et Revocatus leopardum (–os *C apud Spark*) experti . . . vexati sunt *BC ;* i. c. s. revocatus leopardo expertus . . . erat vexatus *AM.*

mis : ita revocatæ et discinctis indutæ. Prior Perpetua iactata est, et cecidit[1] in lumbos. Et ubi sedit, tunicam a latere discissam ad velamentum femoris reduxit, pudoris potius memor quam doloris. Dehinc, acu requisita, et dispersos capillos infibulavit ; non enim decebat martyram sparsis capillis pati, ne in sua gloria plangere videretur. Ita surrexit, et elisam Felicitatem cum vidisset adcessit et manum ei tradidit et suscitavit illam. Et ambæ pariter steterunt, et populi duritia devicta revocatæ sunt in portam Sanavivariam. Illic Perpetua a quodam tunc catechumino, Rustico nomine, qui ei adhaerebat, suscepta, et quasi a somno expergita (adeo in Spiritu et in extasi fuerat) circumspicere cœpit, et instupentibus omnibus ait: Quando, inquit, proicimur[2] ad illam vaccam nescio quam?[3] Et cum audisset quod iam evenerat, non prius credidit, nisi quasdam notas vexationis in corpore et habitu suo recognovisset. Exinde accersitum fratrem suum et illum catechuminum adlocuta est, dicens : In fide state, et invicem omnes diligite ; et passionibus nostris ne scandalizemini.

21. Item Saturus in alia porta Pudentem militem exhortabatur, dicens : Ad summam, inquit, certe sicut præsumpsi et prædixi, nullam usque adhuc bestiam sensi. Et nunc de toto corde credas. Ecce prodeo illo, et ab uno morsu leopardi consumor.[4] Et statim in fine spectaculi, leopardo eiecto, de uno

---

[1] *Mg ;* concidit *A ;* incidit *B.*
[2] *nos ;* proiciemur *M ;* βαλλώμεθα *g ;* producimur *AB.*
[3] quam *BCM ; om. A.*
[4] *ABM ;* τελειοῦμαι (*sc.* consummor) *g.*

morsu tanto perfusus est sanguine ut populus revertenti illi secundi baptismatis testimonium reclamaverit : Salvum lotum, salvum lotum. Plane utique salvus erat, qui hoc modo laverat. Tunc Pudenti militi inquit : Vale, inquit, memento fidei et mei ; et hæc te non conturbent, sed confirment. Simulque ansulam de digito eius petiit, et vulneri suo mersam reddidit ei hereditatem, pignus relinquens illi et memoriam sanguinis.[1] Exinde iam exanimis prosternitur cum ceteris ad iugulationem solito loco. Et cum populus illos in medium postularet, ut gladio penetranti in eorum corpore oculos suos comites homicidii adiungerent, ultro surrexerunt et se quo volebat populus transtulerunt ; ante iam osculati invicem, ut martyrium per sollemnia pacis consummarent. Ceteri quidem inmobiles et cum silentio ferrum receperunt : multo magis Saturus, qui et prior[2] adscenderat, prior reddidit spiritum ; nam et Perpetuam sustinebat. Perpetua autem, ut aliquid doloris gustaret, inter ossa compuncta exululavit ; et errantem dexteram tirunculi gladiatoris ipsa in iugulum suum transtulit. Fortasse tanta femina aliter non potuisset occidi, quæ ab inmundo spiritu timebatur, nisi ipsa voluisset.

O fortissimi ac beatissimi martyres ! O vere vocati et electi in gloriam Domini nostri Iesu Christi ; quam qui magnificat et honorificat et adorat, utique et hæc non minora veteribus exempla

[1] *A ;* tanti sanguinis *BCMg.*
[2] *A ; add.* scalam *BCMg.*

in ædificationem Ecclesiae legere debet, ut novæ quoque virtutes unum et eundem semper Spiritum Sanctum usque adhuc operari testificentur, et [1] omni-potentem Deum Patrem et Filium eius Iesum Christum Dominum nostrum, cui est claritas et inmensa potestas in sæcula sæculorum. Amen.

[1] *Holsten ; codd. alii alia, ut in loco conruptiore.*

THE PASSION OF SS. PERPETUA
AND FELICITY

1. If ancient examples of faith kept, both testi-
fying the grace of God and working the edification
of man, have to this end been set out in writing,
that by their reading as though by the again
showing of the deeds God may be glorified and
man strengthened ; why should not new witnesses
also be so set forth which likewise serve either end ?
Yea, for these things also shall at some time be
ancient and necessary [1] to our sons, though in their
own present time (through some reverence of
antiquity presumed) they are made of but slight
account.   But let those take heed who judge the
one power of one Holy Spirit according to the
succession of times ; whereas those things which
are later ought for their very lateness to be thought
the more eminent, according to the abundance of
grace appointed for the last periods of time.   For
In the last days, saith the Lord, *I will pour forth of
My Spirit upon all flesh, and their sons and their daughters
shall prophesy ; and upon My servants and upon My
handmaids I will pour forth of My Spirit ; and the young
men shall see visions, and the old men shall dream dreams.*
We also therefore, by whom both the prophecies

---

[1] Or perhaps " familiar."

and the new visions promised are received and honoured, and by whom those other wonders of the Holy Spirit are assigned unto the service [1] of the Church, to which also was sent the same Spirit administering all gifts among all men, *according as the Lord hath distributed unto each*—do of necessity both write them and by reading celebrate them to the glory of God ; that no weakness or failing of faith may presume that among those of old time only was the grace of divinity present, whether in martyrs or in revelations vouchsafed ; since God ever works that which He has promised, for a witness to them that believe not and a benefit to them that believe. Wherefore we too, brethren and dear sons, *declare to you* likewise *that which we have heard and handled ;* that both ye who were present may call to mind the glory of the Lord, and ye who now know by hearing may have communion with those holy martyrs, and through them with the Lord Jesus Christ, to Whom is glory and honour for ever and ever. Amen.

2. There were apprehended the young cate-chumens, Revocatus and Felicity his fellow-servant, Saturninus and Secundulus. With them also was Vibia Perpetua, nobly born, reared in a liberal manner, wedded honourably ; having a father and mother and two brothers, one of them a catechumen likewise, and a son, a child at the breast ; and she herself was about twenty-two years of age. What follows here she shall tell herself ; the whole order

[1] Literally. " furnishing."

of her martyrdom as she left it written with her own hand and in her own words.

3. When, saith she, we were yet with our sureties [1] and my father was fain to vex me with his words and continually strove to hurt my faith because of his love : Father, said I, seest thou (for example's sake) this vessel lying, a pitcher or whatsoever it may be ? [2] And he said, I see it. And I said to him, Can it be called by any other name than that which it is ? And he answered, No. So can I call myself nought other than that which I am, a Christian. Then my father moved with this word came upon me to tear out my eyes ; but he vexed me only, and he departed vanquished, he and the arguments of the devil. Then because I was without my father for a few days I gave thanks unto the Lord ; and I was comforted because of his absence. In this same space of a few days we were baptised, and the Spirit declared to me, I must pray for nothing else after that water [3] save only endurance of the flesh. A few days after we were taken into prison, and I was much afraid because I had never known such darkness. O bitter day ! There was a great heat because of the press, there was

[1] See Introduction, p. xiv.
[2] In the symbolism of early Christian art such a vessel was used to signify a Christian's good works, or sometimes the Christian himself as the " vessel of Christ." S. Perpetua may have had this in mind.
[3] The *ab aqua* of the Latin most probably means not " from " the water of baptism, but " after " it ; for the moments just afterwards were held to be specially apt for the request of particular graces (Tert. *De Baptismo*, c. xx.).

cruel handling [1] of the soldiers. Lastly I was tormented there by care for the child. Then Tertius and Pomponius, the blessed deacons who ministered to us, obtained with money that for a few hours we should be taken forth to a better part of the prison and be refreshed. Then all of them going out from the dungeon took their pleasure; I suckled my child that was now faint with hunger. And being careful for him, I spoke to my mother and strengthened my brother and commended my son unto them. I pined because I saw they pined for my sake. Such cares I suffered for many days; and I obtained that the child should abide with me in prison; and straightway I became well, and was lightened of my labour and care for the child; and suddenly the prison was made a palace for me, so that I would sooner be there than anywhere else.

4. Then said my brother to me: Lady my sister, thou art now in high honour, even such that thou mightest ask for a vision; and it should be shown thee whether this be a passion or else a deliverance. And I, as knowing that I conversed with the Lord, for Whose sake I had suffered such things, did promise him, nothing doubting; and I said: To-morrow I will tell thee. And I asked, and this was shown me.

I beheld a ladder of bronze, marvellously great, reaching up to heaven; and it was narrow, so that not more than one might go up at one time. And

[1] Or it may be " extortion."

in the sides of the ladder were planted all manner of things of iron. There were swords there, spears, hooks, and knives ; so that if any that went up took not good heed or looked not upward, he would be torn and his flesh cling to the iron. And there was right at the ladder's foot a serpent [1] lying, marvellously great, which lay in wait for those that would go up, and frightened them that they might not go up. Now Saturus went up first (who afterwards had of his own will given up himself for our sakes, because it was he who had edified us ; and when we were taken he had not been there). And he came to the ladder's head ; and he turned and said : Perpetua, I await thee ; but see that serpent bite thee not. And I said : It shall not hurt me, in the name of Jesus Christ. And from beneath the ladder, as though it feared me, it softly put forth its head ; and as though I trod on the first step I trod on its head. And I went up, and I saw a very great space of garden, and in the midst a man sitting, white-headed, in shepherd's clothing, tall, milking his sheep ; and standing around in white were many thousands. And he raised his head and beheld me and said to me : Welcome, child. And he cried to me, and from the curd he had from the milk he gave me as it were a morsel ; and I took it with joined hands and ate it up ; and all that stood around said, Amen. And at the sound of that word I awoke, yet eating I know not what of sweet.

[1] Or " dragon."

And forthwith I told my brother, and we knew it should be a passion; and we began to have no hope any longer in this world.

5. A few days after, the report went abroad that we were to be tried. Also my father returned from the city spent with weariness; and he came up to me to cast down my faith, saying: Have pity, daughter, on my grey hairs; have pity on thy father, if I am worthy to be called father by thee; if with these hands I have brought thee unto this flower of youth—and I have preferred thee before all thy brothers; give me not over to the reproach of men. Look upon thy brothers; look upon thy mother and mother's sister; look upon thy son, who will not endure to live after thee. Forbear thy resolution; destroy us not all together; for none of us will speak openly among men again if thou sufferest aught. This he said fatherwise in his love, kissing my hands and grovelling at my feet; and with tears he named me, not daughter, but lady. And I was grieved for my father's case because he only would not rejoice at my passion out of all my kin; and I comforted him, saying: That shall be done at this tribunal, whatsoever God shall please; for know that we are not stablished in our own power, but in God's. And he went from me very sorrowful.

6. Another day as we were at meat we were suddenly snatched away to be tried; and we came to the forum. Therewith a report spread abroad through the parts near to the forum, and a very

great multitude gathered together. We went up to the tribunal. The others being asked, confessed. So they came to me. And my father appeared there also, with my son, and would draw me from the step, saying : Sacrifice ; have mercy on the child. And Hilarian the procurator—he that after the death of Minucius Timinian the proconsul had received in his room the right and power of the sword—Spare, said he, thy father's grey hairs ; spare the infancy of the boy. Make sacrifice for the Emperors' [1] prosperity. And I answered : I will not sacrifice. Then said Hilarian : Art thou a Christian ? And I answered : I am a Christian. And when my father stood by me yet to cast down my faith, he was bidden by Hilarian to be cast down and was smitten with a rod. And I sorrowed for my father's harm as though I had been smitten myself ; so sorrowed I for his unhappy old age. Then Hilarian passed sentence upon us all and condemned us to the beasts ; and cheerfully we went down to the dungeon. Then because my child had been wont to take suck of me and to abide with me in the prison, straightway I sent Pomponius the deacon to my father, asking for the child. But my father would not give him. And as God willed, neither is he fain to be suckled any more, nor did I take fever ; that I might not be tormented by care for the child and by the pain of my breasts.

7. A few days after, while we were all praying,

[1] These were the associated Emperors Severus and Caracalla.

suddenly in the midst of the prayer I uttered a word and named Dinocrates ; and I was amazed because he had never come into my mind save then ; and I sorrowed, remembering his fate. And straightway I knew that I was worthy, and that I ought to ask for him. And I began to pray for him long, and to groan unto the Lord. Forthwith the same night, this was shown me.

I beheld Dinocrates coming forth from a dark place, where were many others also ; being both hot and thirsty, his raiment foul, his colour pale ; and the wound on his face which he had when he died. This Dinocrates had been my brother in the flesh, seven years old, who being diseased with ulcers of the face had come to a horrible death, so that his death was abominated of all men. For him therefore I had made my prayer ; and between him and me was a great gulf, so that either might not go to other. There was moreover, in the same place where Dinocrates was, a font full of water, having its edge higher than was the boy's stature ; and Dinocrates stretched up as though to drink. I was sorry that the font had water in it, and yet for the height of the edge he might not drink.

And I awoke, and I knew that my brother was in travail. Yet I was confident I should ease his travail ; and I prayed for him every day till we passed over into the camp prison. (For it was in the camp games that we were to fight ; and the time was the feast of Geta Cæsar.[1]) And I made

---

[1] Geta was the younger brother of Caracalla, and had been

supplication for him day and night with groans and tears, that he might be given me.

8. On the day when we abode in the stocks, this was shown me.

I saw that place which I had before seen, and Dinocrates clean of body, finely clothed, in comfort; and the font I had seen before, the edge of it being drawn down to the boy's navel; and he drew water thence which flowed without ceasing. And on the edge was a golden cup full of water; and Dinocrates came up and began to drink therefrom; which cup failed not. And being satisfied he departed away from the water and began to play as children will, joyfully.

And I awoke. Then I understood that he was translated from his pains.

9. Then a few days after, Pudens the adjutant, in whose charge the prison was, who also began to magnify us because he understood that there was much grace in us, let in many to us that both we and they in turn might be comforted. Now when the day of the games drew near, there came in my father unto me, spent with weariness, and began to pluck out his beard and throw it on the ground and to fall upon his face cursing his years and saying such words as might move all creation. I was grieved for his unhappy old age.

10. The day before we fought, I saw in a vision

raised to the rank of Cæsar in 198. The word *natale* here has not its strict sense of " birthday " (for according to Spartianus, Geta was born on May 27), but means the feast commemorating his elevation.

that Pomponius the deacon had come hither to the door of the prison, and knocked hard upon it. And I went out to him and opened to him ; he was clothed in a white robe ungirdled, having shoes curiously wrought. And he said to me : Perpetua, we await thee ; come. And he took my hand, and we began to go through rugged and winding places. At last with much breathing hard we came to the amphitheatre, and he led me into the midst of the arena. And he said to me : Be not afraid ; I am here with thee and labour together with thee. And he went away. And I saw much people watching closely. And because I knew that I was condemned to the beasts I marvelled that beasts were not sent out against me. And there came out against me a certain ill-favoured Egyptian with his helpers, to fight with me. Also there came to me comely young men, my helpers and aiders. And I was stripped, and I became a man. And my helpers began to rub me with oil as their custom is for a contest ; and over against me I saw that Egyptian wallowing in the dust. And there came forth a man of very great stature, so that he overpassed the very top of the amphitheatre, wearing a robe ungirdled, and beneath it between the two stripes over the breast a robe of purple [1] ; having also shoes curiously wrought in gold and silver ;

---

[1] The interpretation of this difficult passage is probably this. The man wears a white *tunica*, purple-edged and open in front (somewhat like a white cope with purple orphreys) ; beneath it is a purple undergarment, visible between the purple bands of the *tunica*.

bearing a rod like a master of gladiators, and a green branch whereon were golden apples. And he besought silence and said : The Egyptian, if he shall conquer this woman, shall slay her with the sword ; and if she shall conquer him, she shall receive this branch. And he went away. And we came nigh to each other, and began to buffet one another. He was fain to trip up my feet, but I with my heels smote upon his face. And I rose up into the air and began so to smite him as though I trod not the earth. But when I saw that there was yet delay, I joined my hands, setting finger against finger of them. And I caught his head, and he fell upon his face ; and I trod upon his head. And the people began to shout, and my helpers began to sing. And I went up to the master of gladiators and received the branch. And he kissed me and said to me : Daughter, peace be with thee. And I began to go with glory to the gate called the Gate of Life.

And I awoke ; and I understood that I should fight, not with beasts but against the devil ; but I knew that mine was the victory.

Thus far have I written this, till the day before the games ; but the deed of the games themselves let him write who will.

11. And blessed Saturus too delivered this vision which he himself wrote down.

We had suffered, saith he, and we passed out of the flesh, and we began to be carried towards the east by four angels whose hand touched us not.

And we went not as though turned upwards upon our backs, but as though we went up an easy hill. And passing over the world's edge we saw a very great light; and I said to Perpetua (for she was at my side): This is that which the Lord promised us; we have received His promise. And while we were being carried by these same four angels, a great space opened before us, as it had been a pleasure garden, having rose-trees and all kinds of flowers. The height of the trees was after the manner of the cypress, and their leaves sang without ceasing. And there in the garden were four other angels, more glorious than the rest; who when they saw us gave us honour and said to the other angels: Lo, here are they, here are they: and marvelled. And the four angels who bore us set us down trembling; and we passed on foot by a broad way over a plain. There we found Jocundus and Saturninus and Artaxius who in the same persecution had suffered and had been burned alive; and Quintus, a martyr also, who in prison had departed this life; and we asked of them where were the rest. The other angels said to us: Come first, go in, and salute the Lord.

12. And we came near to a place, of which place the walls were such, they seemed built of light; and before the door of that place stood four angels who clothed us when we went in with white raiment. And we went in, and we heard as it were one voice crying *Sanctus, Sanctus, Sanctus* without

any end.[1] And we saw sitting in that same place
as it were a man, white-headed, having hair like
snow, youthful of countenance; whose feet we
saw not. And on his right hand and on his left,
four elders; and behind them stood many other
elders. And we went in with wonder and stood
before the throne; and the four angels raised us
up; and we kissed him, and with his hand he
passed over our faces.[2] And the other elders said
to us: Stand ye. And we stood, and gave the
kiss of peace. And the elders said to us: Go ye
and play. And I said to Perpetua: Thou hast that
which thou desirest. And she said to me: Yea,
God be thanked; so that I that was glad in the
flesh am now more glad.

13. And we went out, and we saw before the
doors, on the right Optatus the bishop, and on the
left Aspasius the priest and teacher, being apart and
sorrowful. And they cast themselves at our feet
and said: Make peace between us, because ye went
forth and left us thus. And we said to them: Art
not thou our Father, and thou our priest, that ye
should throw yourselves at our feet? And we
were moved, and embraced them. And Perpetua
began to talk with them in Greek; and we set
them apart in the pleasure garden beneath a rose
tree. And while we yet spoke with them, the
angels said to them: Let these go and be refreshed;

---

[1] I venture to give the liturgical Latin in my English here; for
the Latin original gives the liturgical Greek: *Agios, agios, agios.*
[2] " And God shall wipe away all tears from their eyes " (Apoc.
VII., 17).

and whatsoever dissensions ye have between you, put them away from you each for each. And they made them to be confounded. And they said to Optatus : Correct thy people ; for they come to thee as those that return from the games and wrangle concerning the parties there. And it seemed to us as though they would shut the gates. And we began to know many brothers there, martyrs also. And we were all sustained there with a savour inexpressible which satisfied us. Then in joy I awoke.

14. These were the glorious visions of those martyrs themselves, the most blessed Saturus and Perpetua, which they themselves wrote down. But Secundulus by an earlier end God called from this world while he was yet in prison ; not without grace, that he should escape the beasts. Yet if not his soul, his flesh at least knew the sword.[1]

15. As for Felicity, she too received this grace of the Lord. For because she was now gone eight months (being indeed with child when she was taken) she was very sorrowful as the day of the games drew near, fearing lest for this cause she should be kept back (for it is not lawful for women that are with child to be brought forth for torment) and lest she should shed her holy and innocent blood after the rest, among strangers and male-factors. Also her fellow martyrs were much afflicted lest they should leave behind them so good

---

[1] The narrator's words are not clear, but seem to mean that Secundulus was beheaded in prison.

a friend and as it were their fellow-traveller on the road of the same hope. Wherefore with joint and united groaning they poured out their prayer to the Lord, three days before the games. Incontinently after their prayer her pains came upon her. And when by reason of the natural difficulty of the eighth month she was oppressed with her travail and made complaint, there said to her one of the servants of the keepers of the door : Thou that thus makest complaint now, what wilt thou do when thou art thrown to the beasts, which thou didst contemn when thou wouldst not sacrifice ? And she answered, I myself now suffer that which I suffer, but there another shall be in me who shall suffer for me, because I am to suffer for him. So she was delivered of a daughter, whom a sister reared up to be her own daughter.

16. Since therefore the Holy Spirit has suffered, and suffering has willed, that the order of the games also should be written ; though we are unworthy to finish the recounting of so great glory, yet we accomplish the will of the most holy Perpetua, nay rather her sacred trust, adding one testimony more of her own steadfastness and height of spirit. When they were being more cruelly handled by the tribune because through advice of certain most despicable men he feared lest by magic charms they might be withdrawn secretly from the prisonhouse, Perpetua answered him to his face : Why dost thou not suffer us to take some comfort, seeing we are victims most noble, namely Cæsar's,

and on his feast day we are to fight? Or is it not thy glory that we should be taken out thither fatter of flesh? The tribune trembled and blushed, and gave order they should be more gently handled, granting that her brothers and the rest should come in and rest with them. Also the adjutant of the prison now believed.

17. Likewise on the day before the games, when at the last feast which they call Free [1] they made (as far as they might) not a Free Feast but a Love Feast, with like hardihood they cast these words at the people; threatening the judgment of the Lord, witnessing to the felicity of their passion, setting at nought the curiosity of those that ran together. And Saturus said: Is not to-morrow sufficient for you? Why do ye favourably behold that which ye hate? Ye are friends to-day, foes to-morrow. Yet mark our faces diligently, that ye may know us again on that day. So they began all to go away thence astonished; of whom many believed.

18. Now dawned the day of their victory, and they went forth from the prison into the amphitheatre as it were into heaven, cheerful and bright of countenance; if they trembled at all, it was for joy, not for fear. Perpetua followed behind, glorious of presence, as a true spouse of Christ and darling of God; at whose piercing look all cast

---

[1] At which ancient, as modern, custom allowed the condemned choice of food and drink. The martyrs used the occasion for the celebration of the *Agape*.

down their eyes. Felicity likewise, rejoicing that
she had borne a child in safety, that she might
fight with the beasts, came now from blood to
blood, from the midwife to the gladiator, to wash
after her travail in a second baptism. And when
they had been brought to the gate and were being
compelled to put on, the men the dress of the
priests of Saturn, the women the dress of the
priestesses of Ceres,[1] the noble Perpetua remained
of like firmness to the end, and would not.[2] For
she said : For this cause came we willingly unto
this, that our liberty might not be obscured. For
this cause have we devoted our lives, that we
might do no such thing as this ; this we agreed
with you. Injustice acknowledged justice ; the
tribune suffered that they should be brought forth
as they were, without more ado. Perpetua began
to sing, as already treading on the Egyptian's head.
Revocatus and Saturninus and Saturus threatened
the people as they gazed. Then when they came
into Hilarian's sight, they began to say to Hilarian,
stretching forth their hands and nodding their
heads : Thou judgest us, said they, and God thee.
At this the people being enraged besought that
they should be vexed with scourges before the line
of gladiators (those namely who fought with beasts).
Then truly they gave thanks because they had received somewhat of the sufferings of the Lord.

[1] The Roman names probably represent the Carthaginian deities
Baal-Ammon and Tanit.
[2] Literally, " that firmness (of hers), noble to the end, resisted."

19. But He Who had said *Ask, and ye shall receive* gave to them asking that end which each had desired. For whenever they spoke together of their desire in their martyrdom, Saturninus for his part would declare that he wished to be thrown to every kind of beast, that so indeed he might wear the more glorious crown. At the beginning of the spectacle therefore himself with Revocatus first had ado with a leopard and was afterwards torn by a bear also upon a raised bridge.[1] Now Saturus detested nothing more than a bear, but was confident already he should die by one bite of a leopard. Therefore when he was being given to a boar, the gladiator instead who had bound him to the boar was torn asunder by the same beast and died after the days of the games ; nor was Saturus more than dragged. Moreover when he had been tied on the bridge to be assaulted by a bear, the bear would not come forth from its den. So Saturus was called back unharmed a second time.

20. But for the women the devil had made ready a most savage cow, prepared for this purpose against all custom ; for even in this beast he would mock their sex. They were stripped therefore and made to put on nets ; and so they were brought forth. The people shuddered, seeing one a tender girl, the other her breasts yet dropping from her

---

[1] See the title-page of Leclercq, *Les Martyrs*, I., for the reproduction of such a scene from a second-century lamp. The " bridge " in the next sentence is distinct from that in this ; it probably crossed the ditch or moat at the edge of the arena.

late childbearing. So they were called back and
clothed in loose robes. Perpetua was first thrown,
and fell upon her loins. And when she had sat
upright, her robe being rent at the side, she drew
it over to cover her thigh, mindful rather of
modesty than of pain. Next, looking for a pin,
she likewise pinned up her dishevelled hair; for
it was not meet that a martyr should suffer with
hair dishevelled, lest she should seem to grieve in
her glory. So she stood up; and when she saw
Felicity smitten down, she went up and gave her
her hand and raised her up. And both of them
stood up together and (the hardness of the people
being now subdued) were called back to the Gate
of Life. There Perpetua being received by one
named Rusticus, then a catechumen, who stood
close at her side, and as now awakening from sleep
(so much was she in the Spirit and in ecstasy) began
first to look about her; and then (which amazed
all there), When, forsooth, quoth she, are we to be
thrown to the cow? And when she heard that
this had been done already, she would not believe
till she perceived some marks of mauling on her
body and on her dress. Thereupon she called her
brother to her, and that catechumen, and spoke to
them, saying : Stand fast in the faith, and love ye
all one another; and be not offended because of
our passion.

21. Saturus also at another gate exhorted Pudens
the soldier, saying : So then indeed, as I trusted
and foretold, I have felt no assault of beasts until

now. And now believe with all thy heart. Behold,
I go out thither and shall perish by one bite of the
leopard. And forthwith at the end of the spectacle,
the leopard being released, with one bite of his he
was covered with so much blood that the people
(in witness to his second baptism) cried out to him
returning : Well washed, well washed. Truly it
was well with him who had washed in this wise.
Then said he to Pudens the soldier : Farewell ;
remember the faith and me ; and let not these things
trouble thee, but strengthen thee. And therewith
he took from Pudens' finger a little ring, and
dipping it in his wound gave it him back again for
an heirloom, leaving him a pledge and memorial
of his blood.[1] Then as the breath left him he was
cast down with the rest in the accustomed place
for his throat to be cut. And when the people
besought that they should be brought forward,
that when the sword pierced through their bodies
their eyes might be joined thereto as witnesses to
the slaughter, they rose of themselves and moved
whither the people willed them, first kissing one
another, that they might accomplish their martyrdom with the rites of peace. The rest not moving
and in silence received the sword ; Saturus much
earlier gave up the ghost ; for he had gone up
earlier also, and now he waited for Perpetua likewise. But Perpetua, that she might have some

---

[1] The end of the soldier's story will be found by those who
wish it in the entry of April 29 in the ancient calendar of the
Church in Carthage : *Pudens Martyr.*

taste of pain, was pierced between the bones and shrieked out; and when the swordsman's hand wandered still (for he was a novice), herself set it upon her own neck. Perchance so great a woman could not else have been slain (being feared of the unclean spirit) had she not herself so willed it.

O most valiant and blessed martyrs! O truly called and elected unto the glory of Our Lord Jesus Christ! Which glory he that magnifies, honours and adores, ought to read these witnesses likewise, as being no less than the old, unto the Church's edification; that these new wonders also may testify that one and the same Holy Spirit works ever until now, and with Him God the Father Almighty, and His Son Jesus Christ Our Lord, to Whom is glory and power unending for ever and ever. Amen.

THE SERMONS OF S. AUGUSTINE UPON THE FEAST OF SS. PERPETUA AND FELICITY

I

To-DAY with its anniversary and return calleth into our mind, and in a manner setteth anew before us, that day whereon the blessed servants of God, Perpetua and Felicity, being adorned with the crowns of martyrdom, did achieve the flower of perpetual felicity; bearing in the battle the name of Christ, and in the prize of battle finding their own. Their exhortations in the heavenly visions, and the triumphs of their passion, we heard when they were read to us; and all these, set out and made clear with the light of words, we have received with our ears, pondered with our minds, honoured with ceremonies of religion, praised with charity. Yet unto so holy a celebration we are bound to give also a solemn homily; and if I that speak it may not set forth their worthiness as I would, yet I bring a ready affection to the joys of so great a feast. For what thing might there be more glorious than these women, whom men may wonder at sooner than they may imitate? But this is chiefly the glory of Him, in Whom they that believe, and they that with holy zeal in His name do contend one with another, are indeed *according to the inward man neither male nor female;* so that even in them that are women in body the manliness

45

of their soul hideth the sex of their flesh, and we
may scarce think of that in their bodily condition
which they suffered not to appear in their deeds.
The dragon therefore was trodden down by the
chaste foot and victorious tread of the blessed
Perpetua, when that upward ladder was shown
her whereby she should go to God ; and the head
of the ancient serpent, which to her that fell was
a stone of stumbling, was made a step unto her
that rose.

What sight may be a more sweet than this, what
strife a more valiant than this, what victory a more
glorious than this ? When their holy bodies were
cast to the beasts, throughout the amphitheatre *the
heathen did rage and the peoples imagine vain things.
But He that dwelleth in Heaven did mock them, and the
Lord laughed them to scorn.* The children of them
whose voices in evil wise raged against the martyrs'
flesh do with godly voices now praise the martyrs'
worth ; nor was the theatre of cruelty then so
filled with them that gathered together unto their
slaughter as is the church of godliness now with
them that gather together unto their honour.
Every year doth charity with religion behold that
which on one day wickedness with sacrilege did
commit. They also beheld, but truly not with the
like intent. They with their cries did that which
the beasts with their biting left yet undone. As for
us, we pity the deeds of the unholy and reverence
the sufferings of the holy. They saw with the eyes
of the flesh that wherewith they might assuage the

46

lust of their hearts ; we with the eyes of our heart
see that which was hidden from them that they
might not see it. They rejoiced over the martyrs
for the death of their bodies ; we sorrow over
themselves for the death of their souls. They
without the light of faith thought the martyrs to
be slain ; we with the strong gaze of faith do
behold them crowned. Lastly, their insulting hath
become our exulting : but this is holy and everlasting ; that was unholy then, and now is nothing.

And for the prizes of martyrs, most beloved, we
believe them to be the chiefest of all ; and rightly
do we believe it. Yet if we diligently consider their
strife, we shall marvel not at the greatness of the
prize. For although this life be toilsome and
fleeting both, yet is there so great a sweetness in it
that albeit men cannot bring it about that they shall
not die, nevertheless they strive much and greatly
that they shall not quickly die. For to banish death
nothing may be done, but to delay death something
may be done. Truly unto every soul toil is weari-
some ; nevertheless even they that look for nothing
after this life, whether good or evil, do toil in all
manner of ways unto this intent, namely that all
their toil be not ended with death. And they that
in error dream of false and carnal delights after
death, and they that with a true faith expect repose
and an unspeakable and blessed rest—do not they
also busy themselves with this and with huge cares
endeavour this, that they shall not quickly die ?
What mean so many labours for the necessity of

food, such enslavement with medicines or other charges (either such as the sick demand or such as is given to them) save that they may not quickly come to that end of death ? At what price then shall be purchased that freedom from death in the life to come, when but the delay thereof in this life hath so great esteem ? For such strange sweetness is in this toilsome life, such is the dread of death in the nature of all that live, howsoever they live, that not even those are content to die who pass through death to the life wherein they may not die.

This joy of living, this fear of dying, the martyrs of Christ with a pure charity, with a certain hope, with a faith unfeigned do by their eminent virtue contemn. In the strength of these they forsake the threats and the promises of the world, they *press forward to the things before.* These trample upon the serpent's head and heed not the manifold hisses of its mouth, but rather rise up thereon. For he hath the victory over all desires who subdueth the tyrannical love of this life whereof all desires are the servants ; nor is a man held by any bond of this life who is not held by the love of life itself. And to the fear of death the pains of the body are wont in some wise to be compared. For sometimes the one, sometimes the other conquereth in a man. A man tormented doth lie that he may not die ; another condemned to death doth lie lest he be tormented. Another speaketh the truth, not because he can bear torment but lest he be tormented if he lie to save himself. But let either fear

soever conquer in any soul soever. The martyrs
of Christ, for the name and the justice of Christ,
won twofold victory; they feared neither to die
nor to suffer pain. He conquered in them Who
lived in them; so that they that lived not unto
themselves but unto Him, in death itself died not.
He showed them His spiritual delights that they
might not feel bodily woes; in such measure as
should suffice not for their failing but for their
trial. In what place was Perpetua, when she felt
not the battle against the maddened cow, when she
asked when that should be done which had been
done already? In what place was she? What saw
she, that she saw not that? What tasted she, that
she felt not that? With what love was she frenzied,
rapt with what sight, drunk with what cup? Nevertheless she clove still to the bonds of the flesh, she
had yet dying members, she was burdened yet with
a corruptible body. What then, when freed from
those bonds and after the pains of that deadly trial
the souls of the martyrs were received and refreshed
with the triumphs of angels; when it was said to
them not: Fulfil that which I have commanded;
but: Receive that which I have promised? With
what joy now do they taste the spiritual banquet?
How rest they in the Lord, in how heavenly a glory
they rejoice, what man with earthly example may
express?

And that life of the blessed martyrs now,
though it passeth already all happiness and delights
of this world, is yet but a little part of the promise,

yea rather a solace of delay. But the day of recompense shall come when every body shall be restored and the whole man shall receive that which he deserveth ; when the limbs of the rich man which once were adorned with a temporal purple shall be tormented with fire everlasting, and the flesh of the poor man that was full of sores shall be changed and shine out amidst the angels ; albeit now also the one in hell thirsteth for a drop of water from the poor man's fingers, and the other in the bosom of the just sweetly reposeth. For even as there is great difference between the joys and sorrows of those that sleep and of those that wake, so also is there great difference between the delights and torments of men that are dead and of men that rise again ; not that the spirits of the dead, as of the sleeping, must needs be deceived, but because the rest of unbodied souls is one, another their glory with heavenly bodies and the felicity of the angels, to whom shall be equalled the host of the faithful that rise again ; among whom the most glorious martyrs shall shine forth with the eminent light of their proper honour, and the same bodies wherein they suffered unworthy torments shall become worthy adornments unto them.

Wherefore, as now we do, let us keep their solemnities with all devotion, with a sober joyfulness, with a holy assembly, with a faithful memory, with believing praise. It is no small part of imitation to rejoice in the virtues of them that are better than we. They are great and we little, but *the Lord hath blessed the little together with the great.* They have gone before us, they have shone out before us. If we may not follow them in deeds, let us follow them in affection ; if not in glory, at least in gladness ; if not in merits, in prayers ; if not in their passion, in our compassion ; if not in eminence, in communion. Let it not seem a little thing to us that we are members of the same body as these to whom we may not be likened. For *if one member suffer, all the members do suffer with it ; so also when one member is* glorified, *all the members rejoice with it.* Glory be to the Head, Who careth both for the hands above and the feet below. As He gave His life for us, one for all, so did these martyrs imitate Him, and gave their lives for their brethren ; and that a fruitful harvest should rise, a harvest of peoples as it were of seeds, they watered the earth with their blood. We also then are the fruits of their labour. We marvel at them, they have compassion on us. We rejoice for them, they pray for us. They strewed their bodies as men their garments when the foal that carried the Lord was led into Jerusalem ; let us at the least cut down branches from the trees, plucking from the sacred Scriptures praises and hymns to bear forth unto the common joy. Yet do we all serve one Lord, follow one Master, attend one King ; we are joined to one Head, journey to one Jerusalem, follow after one charity, embrace one unity.

II

THESE martyrs, brethren, were companions together; but above them all shineth out the name and merit of Perpetua and Felicity, the blessed handmaids of God; for where the sex was more frail, there is the crown more glorious. Truly towards these women a manly courage did work a marvel, when beneath so great a burden their womanly weakness failed not. Well was it for them that they clove unto one husband, even Him unto Whom the Church, being one, is *presented as a chaste virgin*. Well, I say, that they clove to that husband from whom they drew strength to resist the devil; that women should make to fall that enemy who by a woman did make a man to fall. He appeared in them unconquered, Who for their sakes became weak. He filled them with fortitude that He might reap them, Who that He might sow them did empty Himself. He led them unto this glory and honour Who for their sakes did listen to contumely and rebuke. He made these women to die in manly and faithful fashion Who for their sakes did mercifully vouchsafe to be born of a woman.

And it rejoiceth a godly soul to look upon such a sight as the blessed Perpetua hath told was revealed to her of herself, how she became a man and strove with the devil. Truly in that strife she also did run *towards the perfect man, to the measure of the age of the fulness of Christ.* And that ancient and subtle enemy that would leave no device untried, who once by a woman seduced a man and now felt a woman to play the man against him, did strive by a man to vanquish this woman; not without cause. For he set not her husband before her, lest she that by heavenly thoughts already dwelt in the skies, by disdaining suspicion of fleshly love should remain the stronger; but he gave to her father the words of deceit, that the godly soul which might not be softened by the urging of pleasure, might nevertheless by the assault of filial love be broken. In which matter Saint Perpetua answered her father with such temperance that neither did she transgress the commandment which biddeth honour be paid to parents nor yielded to those deceits wherewith that so subtle [1] enemy tried her. And he, being on all sides overcome, caused that same father to be struck with a rod; that whereas she had contemned his words, she might at the least have compassion upon his stripes. And she grieved indeed at that insult upon her aged father, loving him yet to whom she consented not. For she detested the folly in him and not his nature; his infidelity, and not her own birth. Therefore with the greater glory she resisted so beloved a father when he counselled ill, whom she could not see smitten without lamentation; and therefore that sorrow

---

[1] Reading *astutior*, the emendation of the Maurists for MS. *altior*. The Pseudo-Augustine in a parallel passage (Morin, *S. Aug. Tractatus Sive Sermones Inediti*, p. 197, l. 30) has *insidiosius agens*.

took nothing away from the strength of her constancy, but rather it added somewhat to the glory of her passion. For *unto them that love God all things work together for good.*

As for Felicity, she was with child in her very dungeon ; and in her labour did witness unto her woman's lot with a woman's cry. She suffered the pain of Eve, but she tasted the grace of Mary. A woman's debt was required of her, but He succoured her Whom a Virgin bore. Lastly her child was brought forth, timely in an untimely month. For God so willed it that the burden of her womb should not be eased in its rightful time, lest in its rightful time the glory of martyrdom should be delayed. God, I say, so willed it, that the babe should be born out of due season, yet so that to all that company should be given their due Felicity ; lest had she been lacking, there should seem to have lacked not a companion only to the martyrs, but the prize of those same martyrs.[1] For that was the name of these two which is the reward of all. For wherefore do martyrs endure all things if not for this, that they may rejoice in perpetual felicity ? The women therefore were called that unto which all were called. And therefore although there was in that contest a goodly company, with the names of these two the eternity of all is signified, the solemnity of all is sealed.

---

[1] *Ipsorum martyrum præmium.* But perhaps *ipsum* should be read for *ipsorum*—" the very prize of the martyrs."

III

WE keep to-day the feast of those two most holy martyrs who not only in their passion shone out with surpassing virtue but also for that great labour of their piety did seal with their names the reward of themselves and of their comrades likewise. For Perpetua and Felicity are the names of two, but the reward of all. Truly all martyrs would not toil for a while in that strife of passion and confession save that they might rejoice in perpetual felicity. Wherefore by the government of the divine providence it was needful that they should be not martyrs only, but likewise most close companions—as also they were—that they might seal a single day to their glory, and give to them that came after a common solemnity to be kept. For as by the example of their most glorious trial they exhort us unto their imitation, so they testify by their names that we shall receive an inseparable reward. Let both in turn hold it, both weave it together. We hope not for the one without the other. For the perpetual without felicity availeth not, and felicity faileth unless it be perpetual. Now concerning the names of those martyrs to whom this day is dedicate, let these few words suffice.

And for those women whose names these are— even as we heard when their passion was read, and as tradition hath delivered to us and we know, these holy and valiant ones were not only of female kind but were very women. And the one was a mother likewise, that unto the frailty of that sex might be added a more importunate love; so that the Enemy assailing them at all points and hoping they should not bear the bitter and heavy burden of persecution, might think they should straightway yield themselves up to him and be soon his own. But they with the prudent and valiant strength of the inward man did blunt his devices every one and break his assault.

In this company of surpassing glory, men also were martyrs; on that selfsame day most valiant men did suffer and overcome; yet did not they with their names commend this day unto us. And this was so, not because women were preferred before men for the worthiness wherewith they bore themselves, but because the weakness of women more marvellously did vanquish the ancient Enemy, and also the strength of men contended to win a perpetual felicity.

IV

To-day shone forth in the Church two jewels, one brightness; because Perpetua and Felicity both do make one solemnity, nor may any man doubt of that felicity which possesseth a perpetual dignity. They were joined by their custody in the prison, they were joined also by grace; because there is no discord in them. Together they sing in the dungeon, together they go to meet Christ in the air [1]; together they battle against the maddened cow, together they shall enter into their everlasting country; together they suffered their martyrdom; the one suckled her child, the other was in labour. Perpetua said, when she lost her babe and gave up her suckling child: *Who shall separate us from the love of Christ?* Felicity groaned for her labour, and hasted fearlessly after her companions. · And when she was freed from her groaning, what said she to Christ? *Thou hast broken my bonds in sunder; to Thee will I sacrifice the sacrifice of praise.* And the blessed David to comfort her groaning said: The Lord *give thee the desire of thine heart, and strengthen all thine intent.* O frailty! The shadows fled away, but the human estate fled not away. But He that overcame death did deliver her from the peril of childbearing and eased Perpetua from the pain of her breasts. When they climbed the steps of that ladder and trod on the neck of the dragon that lay

---

[1] *Passion*, ch. 12.

in wait,[1] they came to the garden of the celestial meadows and found the good shepherd there who giveth his life for the sheep and seeketh the draught of milk from his flocks. For there sat there, saith she, a shepherd both young and old, fresh in years and hoary of head, who knoweth not age. Youthful was his shining countenance, because he is *ever the same, and his years shall not fail.* Hoary was his head, because in the martyrs the righteous Lord did *love righteousness* and acknowledge equity. Round about him lay his sheep reclining, and himself with a shepherd's hand did milk them in whom he found store of milk and a conscience fruitful of holiness. With his hands he milked them and spoke to them with fatherly comforts, with the heavenly promises that were prepared, saying : *Come, ye blessed of my Father, receive the kingdom which was prepared for you from the foundation of the world.* And he showed them vessels of milk brimming with a pure heart through the shining gift of alms, and said : *I was hungry, and ye gave me to eat ; thirsty, and ye gave me drink.* From this sweet shepherd Perpetua received new milk ere she shed her precious blood. They answered, Amen, and began to ask for the grace of sanctity. They prayed in prison, being now at ease concerning the shepherd. Lord, said they, let not our confession of Thee be dry, that we also may be found worthy to be joined to Thy precious flocks

---

[1] *Passion*, ch. 4. But Felicity was not with Perpetua in the vision ; and in much that follows the writer departs from the original narrative.

and not to be separated from Thy martyrs. And there was set before them in vision a wrestling place, a solemn arena in the amphitheatre. There came that ill-favoured [1] Egyptian who in heaven was the comely Lucifer ; being about to do battle he wallowed in the dust ; and Perpetua being about to triumph in the Lord her Saviour, joined her hands together into a cross, having before her a young man sent by the Lord to defend her. She received from her victory a triumph, she won a branch from her crown. Let us also offer unto them our gifts. Others in that time offered them the visitation of their prison ; let us offer to them the prayers of their solemnity, that with all saints we may be found worthy of a kingdom.

[1] I supply the adjective, which is wanting in the text but is demanded by the antithesis. In the actual vision of Perpetua (ch. 10) the words used are *Ægyptius fœdus specie*. Our author may have repeated *fœdus*, but cadence (and probably rhyme) might suggest a different word. *Lucifer speciosus* would be exactly balanced by *Ægyptius maculosus*.

THE END

DATE DUE

JAN 1 3 1986		
ILL (mail) Luthsion Thsol Ssm 9/25/86		
9/8/8 JUN 1 1989		
JAN 1 9 1993		MAY -7 1994
MAY 2 5 1995	JAN 1 8 1997	
MAR 1 0 1997		
MAY 1 6 1997		
GAYLORD		PRINTED IN U.S.A.

CPSIA information can be obtained
at www.ICGtesting.com
Printed in the USA
BVHW030214270922
648076BV00009B/148